# Technology Textiles
# Volume I

By

Pamela Regan Metts

# Foreword

The Internet of Things (IoT) is where existing things are improved upon with connectivity and data analysis to create intelligence resulting in a decision.

Categories of IoT are:

- Wearable that are worn outside the body
- Home used inside the home
- Medical used for healthcare
- Automotive used in a car or truck.

The IoT category is worth $21 billion which comprises 1.2% of the electronics market. Wearable IoT equals $12 Bn or .7% of the electronic market. In terms of units the prediction is for 295 million wearable, manufactured units to be on the market by 2019 compared to 51 million units of wearable in 2014. (1)

Textiles transcend all categories of IoT. Technology Textiles Volume I outlines some examples of what kind of IoT is coming, what is out there and how they can improve our lives. Also featured in this book are sustainability and social compliance issues applicable to textiles. Other types of innovative and artistic textiles, even some whimsical, are discussed.

## Cupron

Cupron located in Richmond, Virginia has developed copper infused fabric. Copper has been used since Egyptian times for medicinal and cosmetic purposes. Finely ground copper powder is blended with polymeric substances to form a filament to be made into fabric or chips to be molded into hard goods. The copper ions released from substrates using Cupron technology damage microbe cell walls, microbial RNA and DNA and microbial protein in bacterial and fungal cells. Copper infused medical linens and hard surfaces tested at Sentara Leigh

Hospital in 2013 were shown to reduce hospital acquired infection rates by 83% (https://www.sentara.com/hampton-roads-virginia/aboutus/news/copper-clinical-trial.aspx). Up to 90,000 people die a year from hospital acquired infections. Antimicrobials used in Cupron fabric are EPA registered, and Cupron holds 4 public health EPA registrations including registrations to kill athlete's foot fungus on their registered textiles by more than 99.9% after 12 hours of contact.(2)

 Copper infused textiles are a wonderful example of ancient knowledge, present day chemistry and textile technology at work. You can buy product at a variety of retailers, see their website for more details www.cupron.com.

# Sync Footwear

Proto-types of Sync Footwear by H&Y International Technology, Incorporated were on display at the Men's Mart Los Angeles in the Fall of 2016. The shoes flash brightly colored lights in sync with music from a smart phone. The shoe light and color sequence varies with music played on smart phone. The shoes are programmable through an app or by a push button in the shoe. Rechargeable batteries with a USB are in the shoe.

# Conductive Gloves

Conductive gloves knitted with conductivity in the fingertips allow you to use your touch screen devices and keep warm! The gloves were developed by AIQ Smart Clothing. (3)

AIQ was present at the 2016 Wearable Technology Conference in San Francisco; their company representative gave a pitch.

## Waterproof Swimmable

Through innovation and ingenuity, waterproof wearables are on the horizon. Wearable Technologies magazine distributed at the W-T Conference in San Francisco in July of 2016 reported on the following items:

FINIS - Lose the ear buds this device delivers music through your cheekbones to your ears using Bone Conduction Technology. Beethoven first used bone conduction in the 18$^{th}$ century. Beethoven was practically deaf and would bite down on a rod connected to his piano to hear through his jawbone. Soundwaves normally travel though the eardrum which converts the sound waves into vibrations. The vibrations then travel to the inner ear and onto the brain. In bone conduction technology the sound waves are changed to vibrations by the headphone instead of by the eardrum. The vibration then goes through the bone into the inner ear and onto the brain. Bone conduction hearing allows the listener to still hear other sounds through the eardrum allowing the listener to safely navigate through their environment.(4)

ON COURSE - Look at your destination and touch one button to enter course. Green LED in goggle indicates on course, yellow in one eye or other indicates you need to move in opposite direction, red if you're not adjusting correctly. This technology can be very useful in keeping you safe in open water where going off course could result in getting lost and/or becoming overly tired or dehydrated.

INSTABEAT - Heart monitor with color coded LEDs in corner of goggle. Blue indicates a normal heart rate, green indicates optimal exercise beat and red indicates too high a beat. This can keep you safe by making you aware of your heart rate and possibly avoid a cardiac event.

## "Shark Proof" Wetsuits

This is a great example of how printing and technology can work together. Scientific research has shown that certain color patterns are not favourable to sharks. The white and blue camouflage printed Elude suit makes the diver or surfer less visible to the shark due to camouflage that allows wearer to blend in with the blue sea and white foam. The Diverter suit's black and white stripes are intended to look like poisonous sea creatures that sharks tend to avoid.(5)

Light Weight Chain mail offers extra protection from shark attacks. The stainless steel and titanium mesh is becoming lighter and stronger. Neptunic partners with Mailletech Industries in the development of chainmail. (6)

Electro-magnetic suits use conductive carbon fiber in fabric that trap and conceal your naturally-occurring electric signal output that many marine animals can detect and react to.(7)

# Holographic Fabrics

The holographic fabric discussed in this article was developed by me for Barbie in the early 1990s in Los Angeles, in an effort to replace glitter printing.

Holographic fabric is where science meets fashion to create a world of color.
The hologram is created by laser embossing very thin sheets of clear polyester film, vacuum metalizing the film then laminating onto fabric.

The laser embossing creates microscopic grooves on the surface of the clear film.  Light passes through the embossed, clear polyester film and separates it, resulting in an array of the full color spectrum. The arrangement of these grooves creates a holographic image. The clear film is then metalized with a silver aluminium coating to further enhance the hologram's reflective qualities.  The metalizing acts like a mirror to reflect the light.

This coating can be 100% coverage which maximizes the reflective quality. The coating can also be partial, referred to as a percentage of coating, resulting in a transparent, smoky film.

PET (polyester) films are 1/2 mil to 3/4 mil thickness. The embossed, metalized film is then laminated to a substrate of taffeta, tricot or Velcro creating a holographic fabric ready to be cut and sewn.

## Dry Dye

I had the pleasure of visiting the Yeh Group facility in Thailand in 2013 and learned that dry dye is a sustainable method of dyeing synthetic fabrics with carbon dioxide instead of water.

A roll of fabric is installed in a steel chamber similar to those used in deep sea diving. Dyestuff in solid form and $CO_2$ are introduced into the chamber along with extreme pressure.

The dyestuff then sublimates into the roll of fabric dying the entire bolt at one time. If the color is too dark, the roll can be re-introduced into the chamber and dyestuff can be taken out to lighten the color. This technology has the potential to save millions of gallons of water a year used for traditional fabric dying.

## Color Change Fabrics

While working at Mattel in the 1990's and in partnership with Pilot Ink of Japan I developed a variety of color change fabrics for Barbie.

Thermo-chromatic change color when exposed to heat. Photo chromatic colorants change color when exposed to sunlight and hydro chromatic colorants change color when exposed to water. Basically a color change ink changes from color to clear revealing a print or solid color beneath.

Some challenges incurred were cost, ghosting where a dark print or solid might show through a light colored color change ink and partial color change occurring while in transit on very hot boats.

## Tap With Us

TAP prototype on display at the Wearable Technology Conference in July of 2016, indicates that the future of input is at hand.

A wearable Bluetooth keyboard, worn on the hand, turns anything you touch into a typing surface. The hand worn device made of foam is embedded with sensors which accurately detect any combination of finger touches, on any surface and translates into text for texting or writing documents. Learning the tapping sequences to represent the alphabet takes place online.

# Conductive Copper Filament

Copper Filament Powers Up Your Clothes!

As reported in the research journal *Nature Communications* dated 11/11/2016, Associate Professor Jayan Thomas from the University of Central Florida has developed a copper ribbon that can be woven on a loom. The copper ribbon acts as a solar cell on one side and stores energy on the other. The ribbons are so thin so as to resemble and act like a filament that can be woven into fabric. In sunlight, the copper filament can collect energy from the sun and store it. One day your clothes could charge your devices. Professor Thomas was inspired by the movie *Back to the Future*, specifically by Marty McFly's self tying sneakers.

## 30 Year Sustainable Sweatshirt

Designer Tom Cridland of England debuted his 30 year sweatshirt in 2015 to help sustain the environment and provide a quality product to consumers. The sweater comes with 3 decades of free repair and free return postage. (8)

The need for sustainability stems from the fact that 350,000 tons of clothes end up in landfills annually. (9)

360 grams per meter cotton polyester fabric and double reinforced stitching help ensure the durability of the sweatshirt. (10)

Sustainability is the key to this long-use sweater unlike some garments developed with planned obsolescence in mind.

## Smart Jacket

Due out Fall of 2017 the smart denim trucker jacket is touch sensitive and connected to your device. The jacket uses conductive, washable threads woven into the jacket fabric. A detachable cuff piece contains other components creating a live jacket to safely use a cell phone without actually touching it. By swiping touch- sensitive areas on the jacket you can access directions, music, phone calls or even where to stop for coffee safely while riding a bicycle. The touch sensitive textiles are made possible using:

- Detachable smart tag that houses electronics:
    - Connectors
    - Tiny circuits
    - All fits in the size of a button
- Conductive Yarns:
    - Custom made touch and gesture sensitive locations
    - Sensor grids create big interactive areas

- Yarns contain metal alloys with natural and manmade yarns like cotton, polyester or silk.
- Looks and feels like regular yarn

- Woven Fabrics:
  - Can be woven on factory looms
  - Denim and many other weaves
  - Washable
  - Soft and pliable

The collaborative effort between bay area Google and Levi's is known as Project Jacquard.(11)

## Buffalo Fibers

Buffalo fiber is picked up off the ground as the buffalo naturally shed, off fences using street sweeper brooms, harvested at meat packing plants and sometimes cautiously sheared at hydraulic chutes during round-up. Buffalo are indigenous to the United States.

The Woolery, United by Blue and The Buffalo Wool Company provide Buffalo fiber products such as gloves, socks, and much more. The Buffalo Wool Company began developing the buffalo fiber around 2007.

Buffalo fibers have following attributes:

- Warmth
- Lightweight
- Springy
- Hypo-Allergenic

- Anti-Microbial

- Easy Care and Durable

- Native to America(12)

## Know the Chain

21 million people have becomes victims of forced labor resulting in $150 billion dollars of illegal profits in the private sector alone.(13) The California Supply Chain Transparency Act was passed into law in 2010 to help combat forced labor. It is a disclosure law requiring retailers and manufacturers to have prominently displayed on their websites specific facts regarding labor law compliance. The disclosure encompasses audits, supplier certifications, internal accountability and training.(14)

Forced labor is often an organized crime using agents. Human trafficking is prominent in certain parts of the globe and the victims are often recruited by nefarious agents who pose as legitimate employment agencies, preying on weak and isolated people.

Visit KnowtheChain.org to see a list of organizations and how they rank in the handling of forced labor. Compiled in 2016, on a global basis, the list serves to bring to light the issue of forced labor.

## Spider Yarn

Adidas and Biosteel developed a new biodegradable shoe in 2017 made from Spider Yarn. The shoe, after about 2 years of use, can be dissolved (except for foam sole) in your sink using a protein enzyme. (15)

Spider silk is made from the protein-based, bio-polymer fiber Biosteel from the German company AMSilk. (16)

Meanwhile, back in the lab, Biocraft Lab scientists, in partnership with University of Notre Dame, are injecting silk worm embryos with spider DNA using a micro-injection apparatus. 10 out of 500 silk worm moths can lay eggs with spider DNA that spin their cacoons with extra strong filament. The filament is harvested from the cacoons. (17)

## Embroline

Swiss company Coloreel unveiled its thread coloring embroidery machine attachment Embroline in early 2017 at the Avantex tradeshow in Paris.

The thread coloring attachment is based on inkjet technology, similar to that used in digital textile printing. The thread coloring attachment can be used on practically any embroidery machine and colors the Embroline base thread instantly using Coloreel approved inks. Embroline coloring software enables colors to be changed from one color to another or gradually change color, creating a gradient effect. The patented Embroline coloring attachment streamlines the embroidery process by as much as 80% by eliminating the need to cut threads and change reels with each color change. Thread is colored instantly from one single reel while being embroidered. Embroline embroidery does not require post treatment. Embroline also lends itself to sustainability by eliminating the need to keep stock of multiple thread colors. (18)

# Elbow Flexion

The Elbow Flexion is a flexible sleeve, not a brace, that uses sewn in sensors and power sources to wirelessly transmit tracking data to an off-site doctor to monitor the healing process and reduce recovery time for elbow injuries. Textile based sensors and textile based power generation methods are the elements behind Ohio State University junior Raman Vilkhu's Elbow Flexion joint monitoring sleeve. The wireless sleeve project won Electrical and Computer engineering student, Raman Vilkhu, the IEEE Microwave Theory and Techniques (MTT-S) Spring 2017 Scholarship award. The MTT-S promotes the advancement of microwave theory and its applications including RF, microwave, and millimetre wave and terahertz technologies.(19)

# Kniterature

On display at The 14th Factory Gallery in April 2017 in Los Angeles, Kniterature is 15 meters of knitted magazines and books by artist Movanna Chen.  Movanna learned how to knit as a child in order to send knit goods to her brothers and sisters away at school.  She became an accountant and one day while shredding documents and reflecting on her training in fashion, thought of trying to knit using the shredded paper goods.  The shredded paper has to be knitted using several plies unlike wool that can be knitted using a single ply yarn.  Movanna has knitted dresses, enclosed herself in her creations and stood in public places to see what would happen.  Many people looked, a few, particularly older people asked her what she was doing.  She has also brought groups of people together, taught them how to knit, then let the conversation flow.  Kniterature is way to generate conversation and look at how we view fashion and the media in a new light.(20)

## Cat Watches

Pet and dog fur watches are being offered for sale by the Analogwatchco.com. Just send in 2-4 oz of your pet's fur in an airtight bag, the fur is hand crafted into a felted wool-like substrate, which is heat formed to a leather watch band and metal watch body shell. With a hydrophobic coating, the timepiece will remain soft, yet stay clean and water resistant. (21)

## Power Up with Diamonds!

Michigan State University Fraunhofer Center are growing diamonds! Michigan State University in Lansing, Michigan and Fraunhofer Center for Coatings and Diamond Technology of Dresden, Germany funded a $5 million dollar expansion of a diamond growing laboratory in Lansing, Michigan in 2015.

Diamonds are being grown in layers using diamond synthesis equipment. Coatings are also being developed to reduce friction between moving parts. Diamonds possess many important properties that make them good at distributing heat; poor distribution of heat account for many electronic failures. (22)

## Environmental Profit and Loss

Kering Group home of luxury brands such as Gucci, Stella McCarthy and sportswear brand Volcom developed its Environmental Profit and Loss in 2015 as a way to improve the environment and use limited resources more wisely. Kering Group started in the 1960s had gross revenues of 12.4 billion Euros in 2016 and employed more than 40, 0000 people. Based in France, Kering Group is a global company.

The EP&L developed by Kering Group according to their website Kering.com is a "management tool that measures and monetizes the environmental impacts of activities..." Activities includes: sourcing, production and supplier performance.

The EP&L follows these steps:
- Decide what to measure
- Map the supply chain
- Identify priority data
- Collect primary data - internal data
- Collect secondary data - external data
- Determine valuation - monetize
- Calculate and analyze results

One result of the EP&L for Kering Group was their brand Bottega Veneta was able to purchase 54,000 square meters of leather tanned without the use of metals or chrome. (23)

# Epilogue

Please go to my website **www.technologytextiles.com** where there are live links to learn even more about these articles.

Technology Textiles Volume I is a culmination and expansion of articles from TechnologyTextiles.com written between July 2016 and May 2017. TechnologyTextiles.com is a website I started in 2016 dedicated to pushing textiles foreword through technology. I am a graduate of Philadelphia University, with a bachelor of science in Textile Management and Marketing and a graduate of California State University Dominguez Hills' MBA program. I have over 20 years experience including close to 15 years as a Textile Engineer at Mattel, three years as a consultant Textile Engineer at Spinmaster, and over three years as a Regulatory Compliance Manager at VF Corp. My work has taken me throughout the United States, China, Hong Kong, Korea, Japan, Taiwan and Thailand.

## Citations

1. "IoT Roadmap 2017," by Satish Parupalli, May 22, 2017, at **https://global.gotowebinar.com/join/6113460954603918082/494176955**

2. "Copper Based Antimicrobial Technology." Cupron. Accessed May 20, 2017. **https://cupron.com/**.

3. "AIQ: Smart Clothing." AIQ: Smart Clothing. Accessed May 22, 2017. **http://www.aiqsmartclothing.com/**.

4. "Bone conduction waterproof headphones for sports, running and swimming." Audio Bone Headphones. Accessed May 22, 2017. **http://www.audioboneheadphones.com/**.

5. "Australian firm develops 'shark-proof' wetsuits." BBC News. July 18, 2013. Accessed May 22, 2017. **http://www.bbc.com/news/world-asia-23357682**.

6. "Neptunic Sharksuits." Neptunic. Accessed May 22, 2017. **https://neptunic.com/**.

7. "Ask an Expert about SharkSkin Covert Electromagnetic Stinger Front-Zip Wetsuit at Scuba.com." Scuba Gear at Scuba.com. Accessed May 22, 2017. http://www.scuba.com/shop/AskAnExpert.aspx?sku=154496.

8. "The World's Number 1 Sustainable Fashion Brand." Tom Cridland. Accessed May 23, 2017. https://www.tomcridland.com/.

9. Reporter, Louisa Pilbeam Sky News. "Eco-Fashion: Demand for greener clothes grows." Sky News. May 07, 2017. Accessed May 23, 2017. **http://news.sky.com/story/future-of-fashion-demand-for-greener-clothes-grows-10866848**.

10. Seo, Juyoung. "Sustainable Luxury: Tom Cridland's "30 Year Sweatshirt"." Forbes. March 11, 2016. Accessed May 23, 2017. https://www.forbes.com/sites/juyoungseo/2016/03/10/sustainable-luxury-tom-cridlands-30-year-sweatshirt/.

11. "Project Jacquard." Google. Accessed May 23, 2017. **https://atap.google.com/jacquard/**.

12. Co., The Buffalo Wool. "Why Bison?" The Buffalo Wool Co. Accessed May 23,

2017. https://thebuffalowoolco.com/pages/why-bison.

13. "KnowTheChain." KnowTheChain. Accessed May 23, 2017. https://knowthechain.org/.

14. "The California Transparency in Supply Chains Act - oag.ca.gov." Accessed May 23, 2017. https://www.bing.com/cr?IG=F25B65652168442C9E10BCABD9940CA3&CID=2F155E50E0176C0A0CB654DBE1116DC8&rd=1&h=cE0a1xMQojf4ndumubh68Zlzs3kurK-ahGmr2zyXD-4&v=1&r=https%3a%2f%2foag.ca.gov%2fsites%2fall%2ffiles%2fagweb%2fpdfs%2fsb657%2fresource-guide.pdf&p=DevEx,5062.1.

15. "Adidas Official Website | adidas." Adidas United States. Accessed May 24, 2017. http://www.adidas.com/us/.

16. "Get more information about our Biosteel® fiber." Biosteel Fibers | AMSilk. Accessed May 24, 2017. https://www.amsilk.com/industries/biosteel-fibers/#c156.

17. "Exploration Earth 2050." In *Exploration Earth 2050*. Fox. April 1, 2017.

18. "Coloreel - Revolutionary thread colouring." Embroline by Coloreel. April 05, 2017. Accessed May 24, 2017. http://www.coloreel.com/.

19. "ECE undergrad wins scholarship for wearable electronic joint monitor." Electrical and Computer Engineering. Accessed May 24, 2017. https://ece.osu.edu/news/2017/02/ece-undergrad-wins-scholarship-wearable-electronic-joint-monitor.

20. "Movana chen." Movana chen. Accessed May 24, 2017. http://www.movanachen.com/.

21. "THE COMPANION COLLECTION." Analog Watch Co. Accessed May 24, 2017. https://analogwatchco.com/products/the-companion-collection?variant=40296996167.

22. "Diamond for Power Electronics." Diamond for Power Electronics | Fraunhofer Center for Coatings and Diamond Technologies (CCD). Accessed

May 24, 2017.
https://www.egr.msu.edu/fraunhofer-ccd/projects/diamond-power-electronics.

23. "Sustainability." Sustainability | Kering. Accessed May 24, 2017. http://www.kering.com/en/sustainability.

## Bibliography

"Adidas Official Website | adidas." Adidas United States. Accessed May 24, 2017. **http://www.adidas.com/us/**.

"AIQ: Smart Clothing." AIQ: Smart Clothing. Accessed May 22, 2017. http://www.aiqsmartclothing.com/.

"Ask an Expert about SharkSkin Covert Electromagnetic Stinger Front-Zip Wetsuit at Scuba.com." Scuba Gear at Scuba.com. Accessed May 22, 2017. **http://www.scuba.com/shop/AskAnExpert.aspx?sku=154496**.

"Australian firm develops 'shark-proof' wetsuits." BBC News. July 18, 2013. Accessed May 22, 2017. **http://www.bbc.com/news/world-asia-23357682**

"Bone conduction waterproof headphones for sports, running and swimming." Audio Bone Headphones. Accessed May 22, 2017. **http://www.audioboneheadphones.com/**. Co., The Buffalo Wool. "Why Bison?" The Buffalo Wool Co. Accessed May 23, 2017. **https://thebuffalowoolco.com/pages/why-bison**.

"Coloreel - Revolutionary thread colouring." Embroline by Coloreel. April 05, 2017. Accessed May 24, 2017. **http://www.coloreel.com/**.

"Copper Based Antimicrobial Technology." Cupron. Accessed May 20, 2017. **https://cupron.com/**.

"Diamond for Power Electronics." Diamond for Power Electronics | Fraunhofer Center for Coatings and Diamond Technologies (CCD). Accessed May 24, 2017. https://www.egr.msu.edu/fraunhofer-ccd/projects/diamond-power-electronics.

"ECE undergrad wins scholarship for wearable electronic joint monitor." Electrical and Computer Engineering. Accessed May 24, 2017. https://ece.osu.edu/news/2017/02/ece-undergrad-wins-scholarship-wearable-electronic-joint-monitor.

"Exploration Earth 2050." In *Exploration Earth 2050*. Fox. April 1, 2017.

"Get more information about our Biosteel® fiber." Biosteel Fibers | AMSilk. Accessed May 24, 2017. https://www.amsilk.com/industries/biosteel-fibers/#c156.

"KnowTheChain." KnowTheChain. Accessed May 23, 2017. https://knowthechain.org/

"Movana chen." Movana chen. Accessed May 24, 2017. http://www.movanachen.com/.

"Neptunic Sharksuits." Neptunic. Accessed May 22, 2017. https://neptunic.com/.

"Project Jacquard." Google. Accessed May 23, 2017. https://atap.google.com/jacquard/.

"Ramandeep Vilkhu Receives Prestigious Undergraduate Scholarship Award." ElectroScience Laboratory. Accessed May 24, 2017. https://electroscience.osu.edu/news/2017/02/ramandeep-vilkhu-receives-prestigious-undergraduate-scholarship-award.

Reporter, Louisa Pilbeam Sky News. "Eco-Fashion: Demand for greener clothes grows." Sky News. May 07, 2017. Accessed May 23, 2017. http://news.sky.com/story/future-of-fashion-demand-for-greener-clothes-grows-10866848.

Seo, Juyoung. "Sustainable Luxury: Tom Cridland's "30 Year Sweatshirt"." Forbes. March 11, 2016. Accessed May 23, 2017. https://www.forbes.com/sites/juyoungseo/2016/03/10/sustainable-luxury-tom-cridlands-30-year-sweatshirt/.

"Sharkproof Wetsuits." Neptunic. Accessed May 22, 2017. https://neptunic.com/.
"Sustainability." Sustainability | Kering. Accessed May 24, 2017. http://www.kering.com/en/sustainability.

"The California Transparency in Supply Chains Act - oag.ca.gov." Accessed May 23, 2017. https://www.bing.com/cr?IG=F25B65652168442C9E10BCABD9940CA3&CID=2F155E50E0176C0A0CB654DBE1116DC8&rd=1&h=cE0a1xMQojf4ndumubh68Zlzs3kurK-ahGmr2zyXD-4&v=1&r=https%3a%2f%2foag.ca.gov%2fsites%2fall%2ffiles%2fagweb%2fpdfs%2fsb657%2fresource-guide.pdf&p=DevEx,5062.1.

"THE COMPANION COLLECTION." Analog Watch Co. Accessed May 24, 2017. https://analogwatchco.com/products/the-companion-collection?variant=40296996167.

"The World's Number 1 Sustainable Fashion Brand." Tom Cridland. Accessed May 23, 2017. https://www.tomcridland.com/.

www.ingramcontent.com/pod-product-compliance
Lightning Source LLC
Chambersburg PA
CBHW020957180526
45163CB00006B/2405